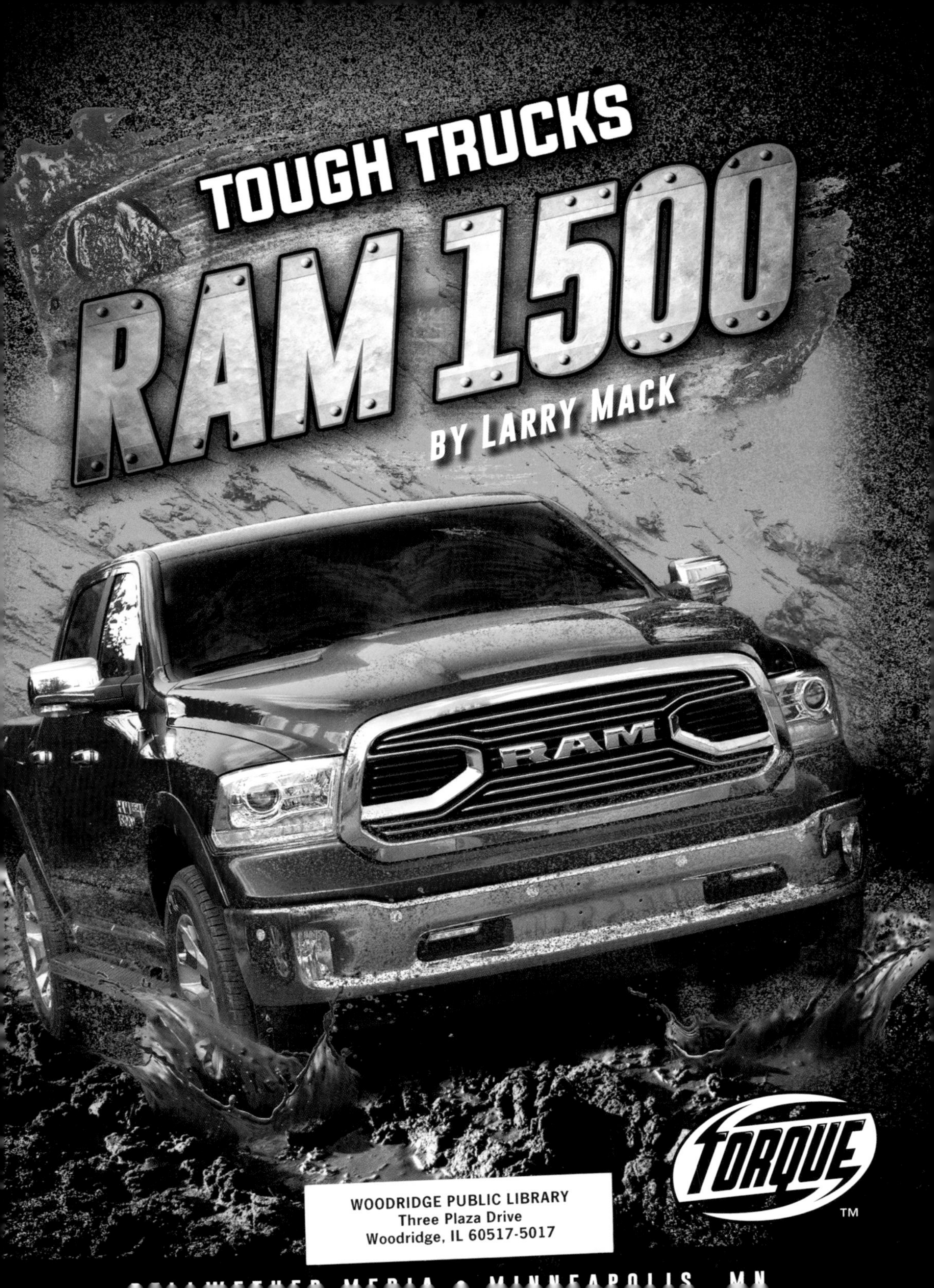

BELLWETHER MEDIA • MINNEAPOLIS, MN

Are you ready to take it to the extreme? Torque books thrust you into the action-packed world of sports, vehicles, mystery, and adventure. These books may include dirt, smoke, fire, and dangerous stunts.
WARNING read at your own risk.

This edition first published in 2019 by Bellwether Media, Inc.

Library of Congress Cataloging-in-Publication Data

Names: Mack, Larry, author.
Title: Ram 1500 / by Larry Mack.
Description: Minneapolis, MN : Bellwether Media, Inc., 2019. | Series:
Torque: Tough Trucks | Includes bibliographical references and index. |
Audience: Ages 7-12.
Identifiers: LCCN 2018002183 (print) | LCCN 2018006817 (ebook) | ISBN
9781626178953 (hardcover : alk. paper)| ISBN 9781681036144 (ebook)
Subjects: LCSH: Ram truck–Juvenile literature.
Classification: LCC TL230.5.R35 (ebook) | LCC TL230.5.R35 M33 2019 (print) |
DDC 629.223/2-dc23
LC record available at https://lccn.loc.gov/2018002183

Editor: Betsy Rathburn Designer: Josh Brink

Printed in the United States of America, North Mankato, MN.

TABLE OF CONTENTS

A RAM IN THE SNOW

The air is cold and clean. A fresh blanket of snow covers the ground. A Ram 1500 pickup truck makes its way up a high mountain road.

Inside the **cab**, the driver is warm and comfortable. He steers the truck along the road's twists and curves. There are just a few more miles to his warm mountain cabin.

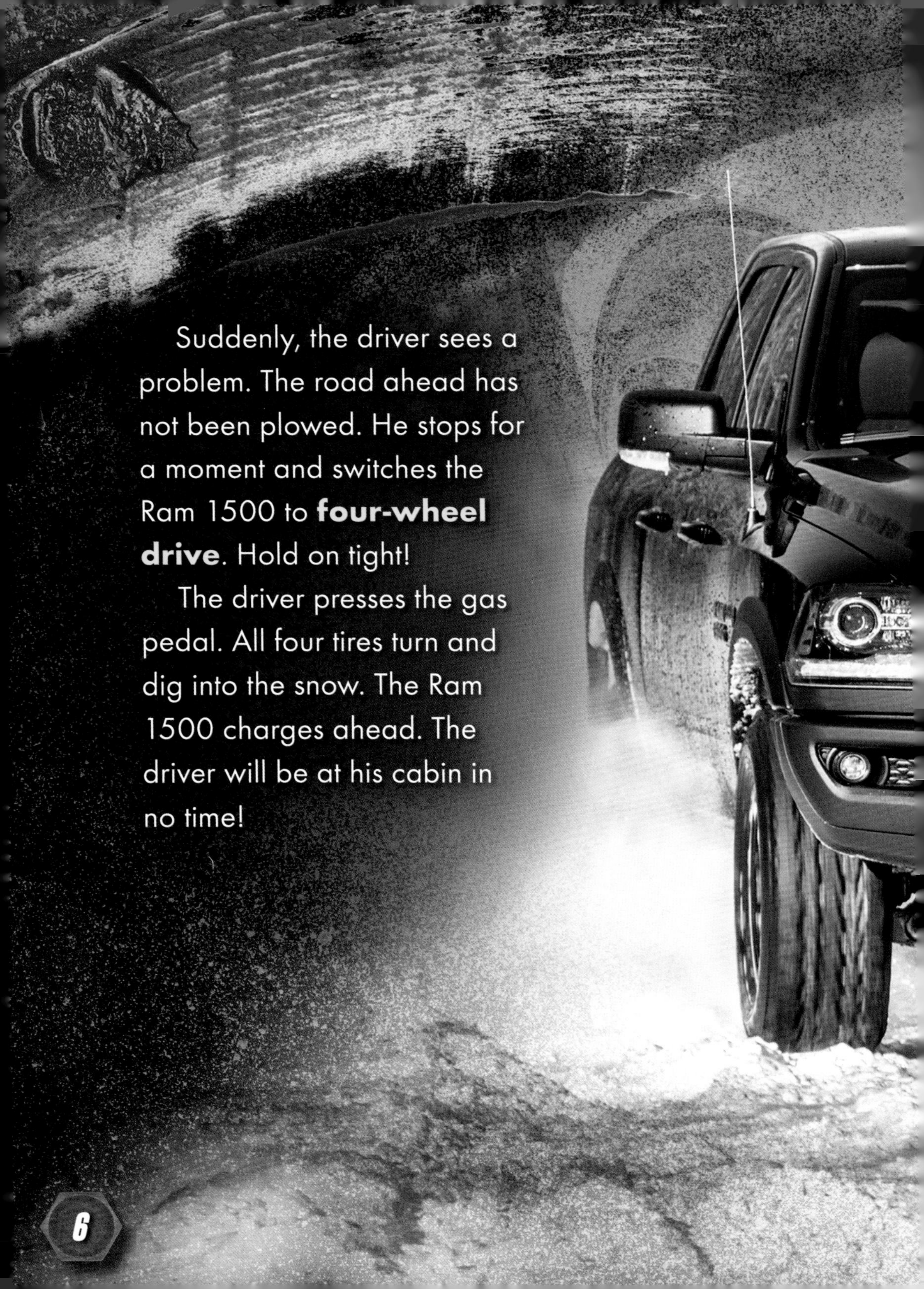

Suddenly, the driver sees a problem. The road ahead has not been plowed. He stops for a moment and switches the Ram 1500 to **four-wheel drive**. Hold on tight!

The driver presses the gas pedal. All four tires turn and dig into the snow. The Ram 1500 charges ahead. The driver will be at his cabin in no time!

RAM

RAM 1500 HISTORY

The Ram 1500's history began with two brothers. John and Horace Dodge built their first car in 1914. The company they created would grow to become one of the most famous American car companies.

Horace Dodge

John Dodge

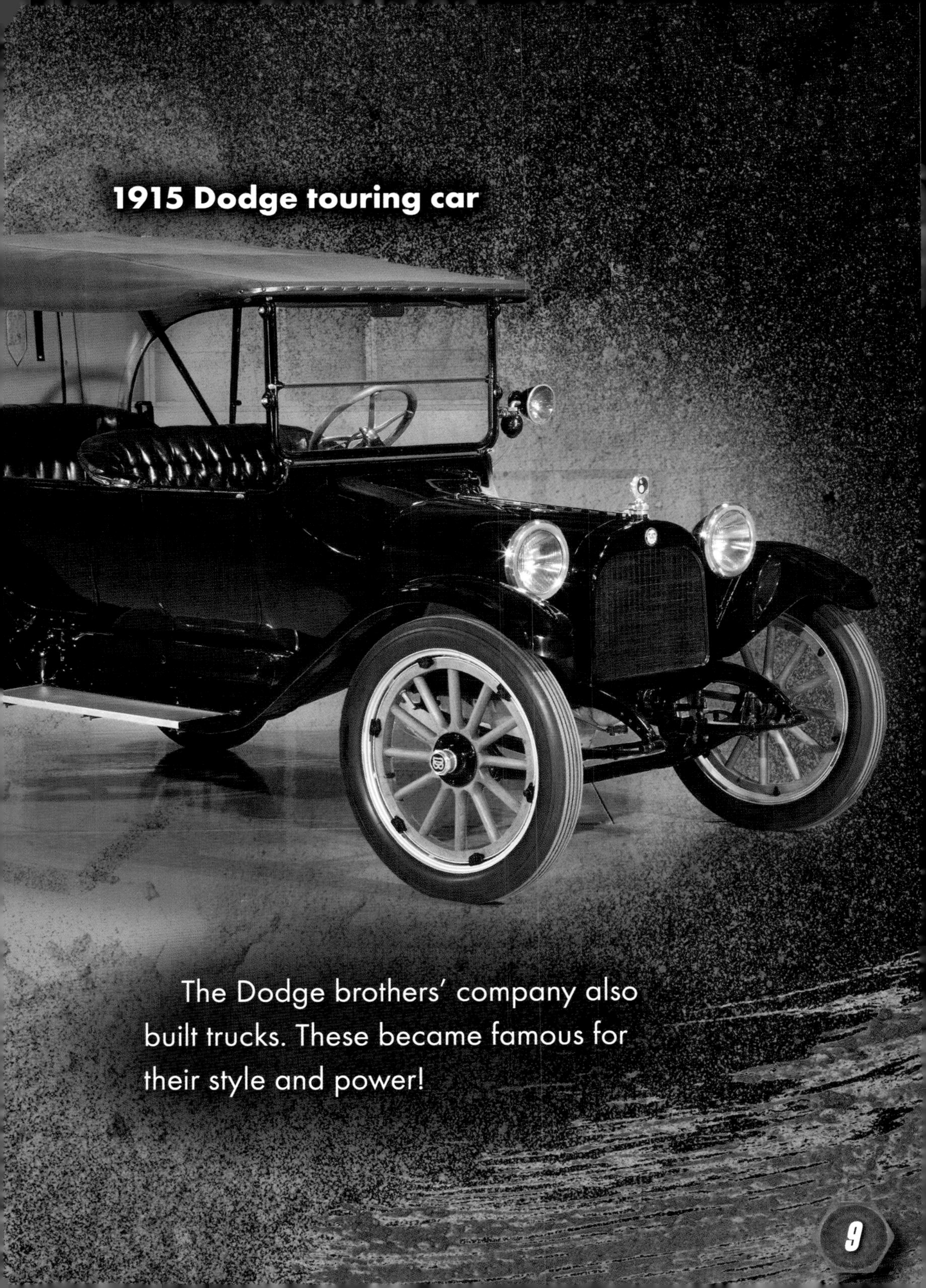

1915 Dodge touring car

The Dodge brothers' company also built trucks. These became famous for their style and power!

early 1980s Dodge Ram

Early Dodge trucks had a ram **hood ornament** to show that they were tough like the animal. Later, Dodge dropped the ram symbol. It reappeared in 1981 when the company introduced new Ram trucks.

RAM TOUGH

THE RAM IS A MEMBER OF THE SHEEP FAMILY. WHEN RAMS FIGHT, THEY CHARGE EACH OTHER AND KNOCK HEADS AT ABOUT 20 MILES (32 KILOMETERS) PER HOUR!

hood ornament

In 2009, there was an even bigger change at Dodge. The Ram truck **brand** split from the Dodge company. Now, the Ram Truck **Division** builds and sells only trucks!

RAM 1500 TODAY

The Ram Trucks Division has paid off. Ram has won awards from top magazines like *Motor Trend*. Some Ram 1500s have been named Truck of the Year!

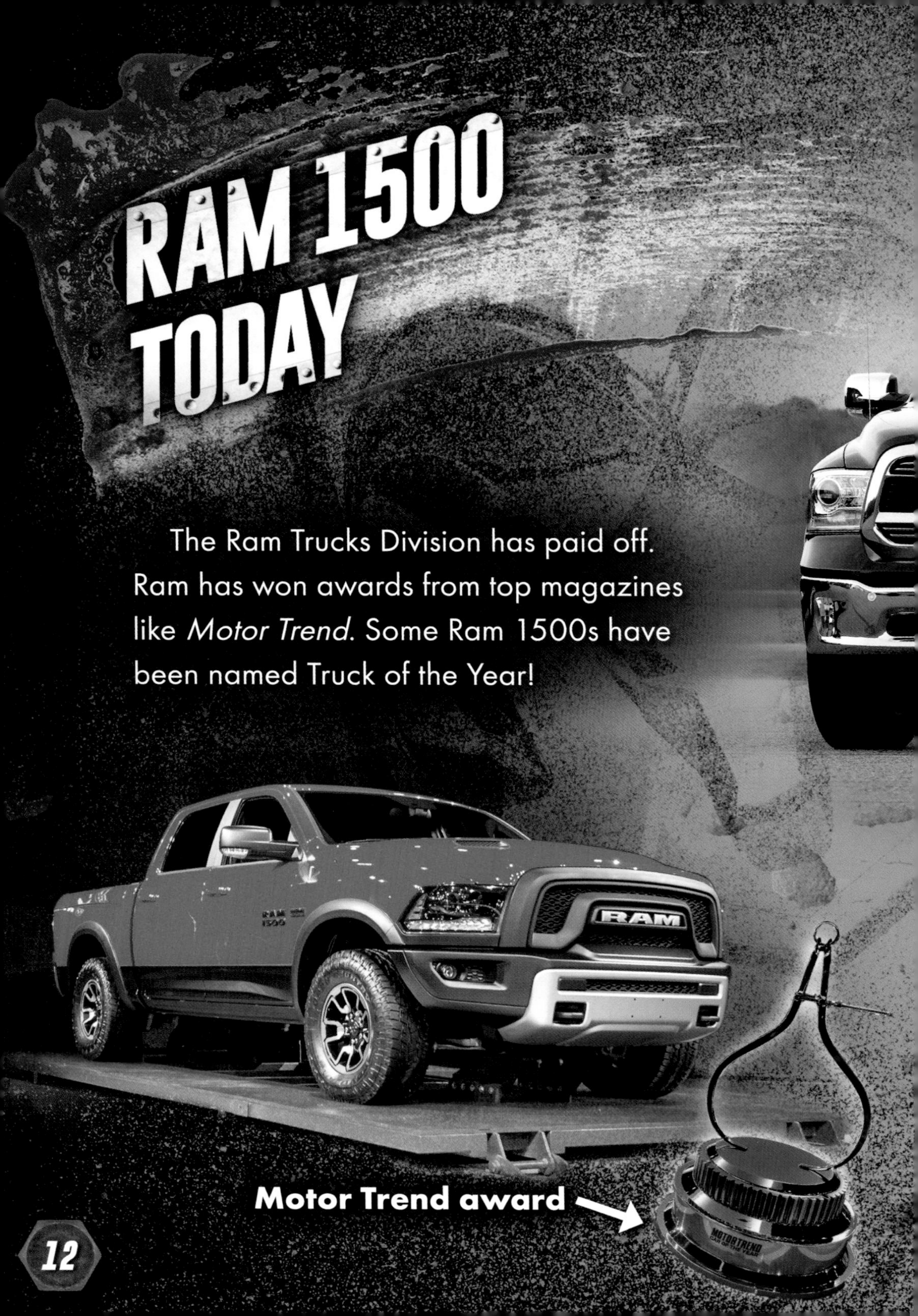

Motor Trend award

The Ram 1500 is popular among pickup fans. It comes in many versions. Each has great towing and hauling ability. The most powerful Ram 1500s can pull more than 10,000 pounds (4,536 kilograms)!

FEATURES AND TECHNOLOGY

Pickups are designed for hauling and towing. The power a truck needs for these jobs comes from its engine. Ram trucks have powerful engines with six or eight **cylinders**.

1951 Chrysler New Yorker

HEMI HISTORY

CHRYSLER BOUGHT THE DODGE COMPANY IN 1928. THEN IN 1951, THE COMPANY RELEASED THE FIRST CARS WITH HEMI ENGINES!

Any Ram can be bought with the famous HEMI **V8 engine**. HEMI engines have been used in **muscle cars** and race cars!

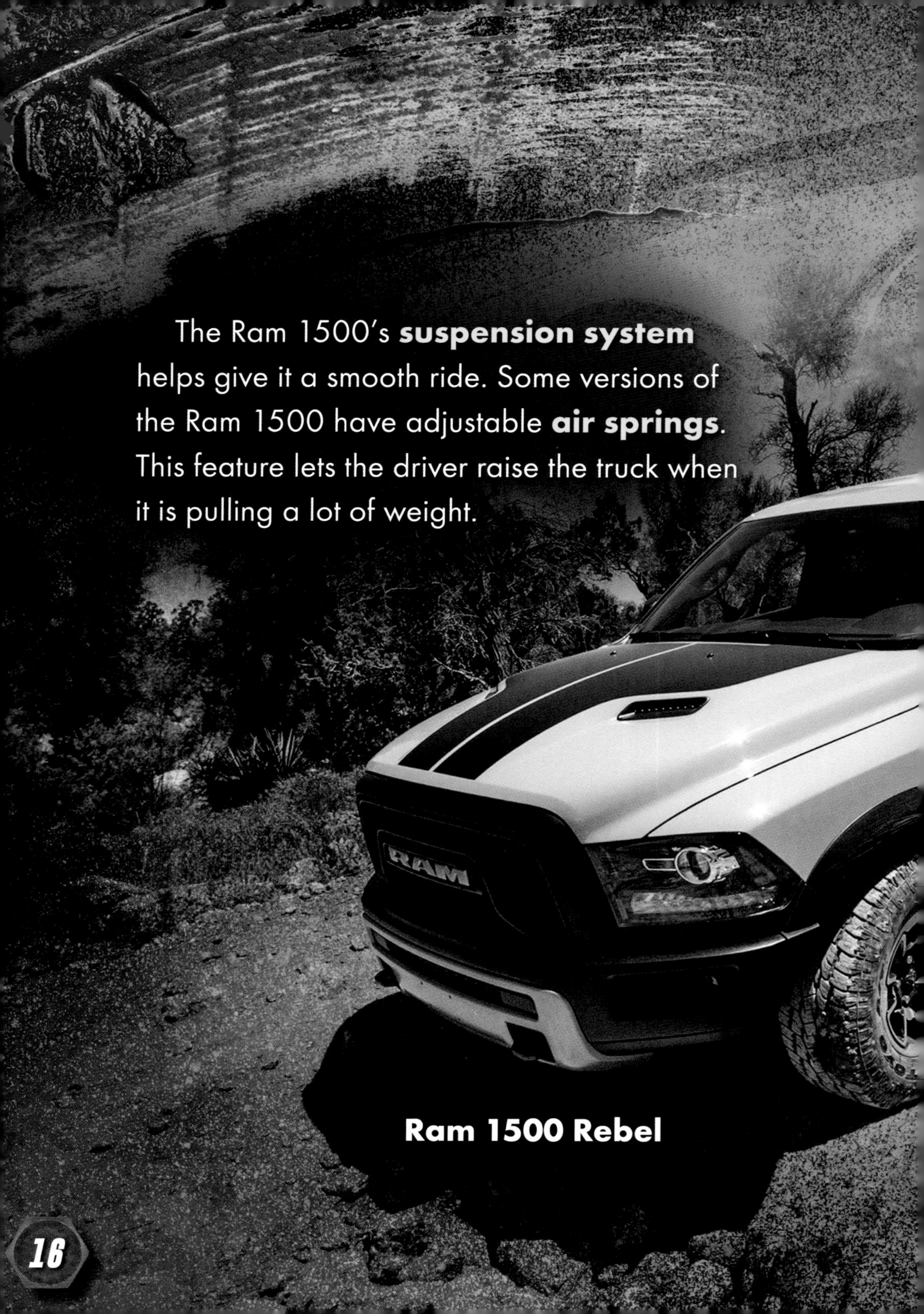

The Ram 1500's **suspension system** helps give it a smooth ride. Some versions of the Ram 1500 have adjustable **air springs**. This feature lets the driver raise the truck when it is pulling a lot of weight.

Ram 1500 Rebel

The springs lower the truck, too. It automatically lowers when it is going fast. This helps save fuel!

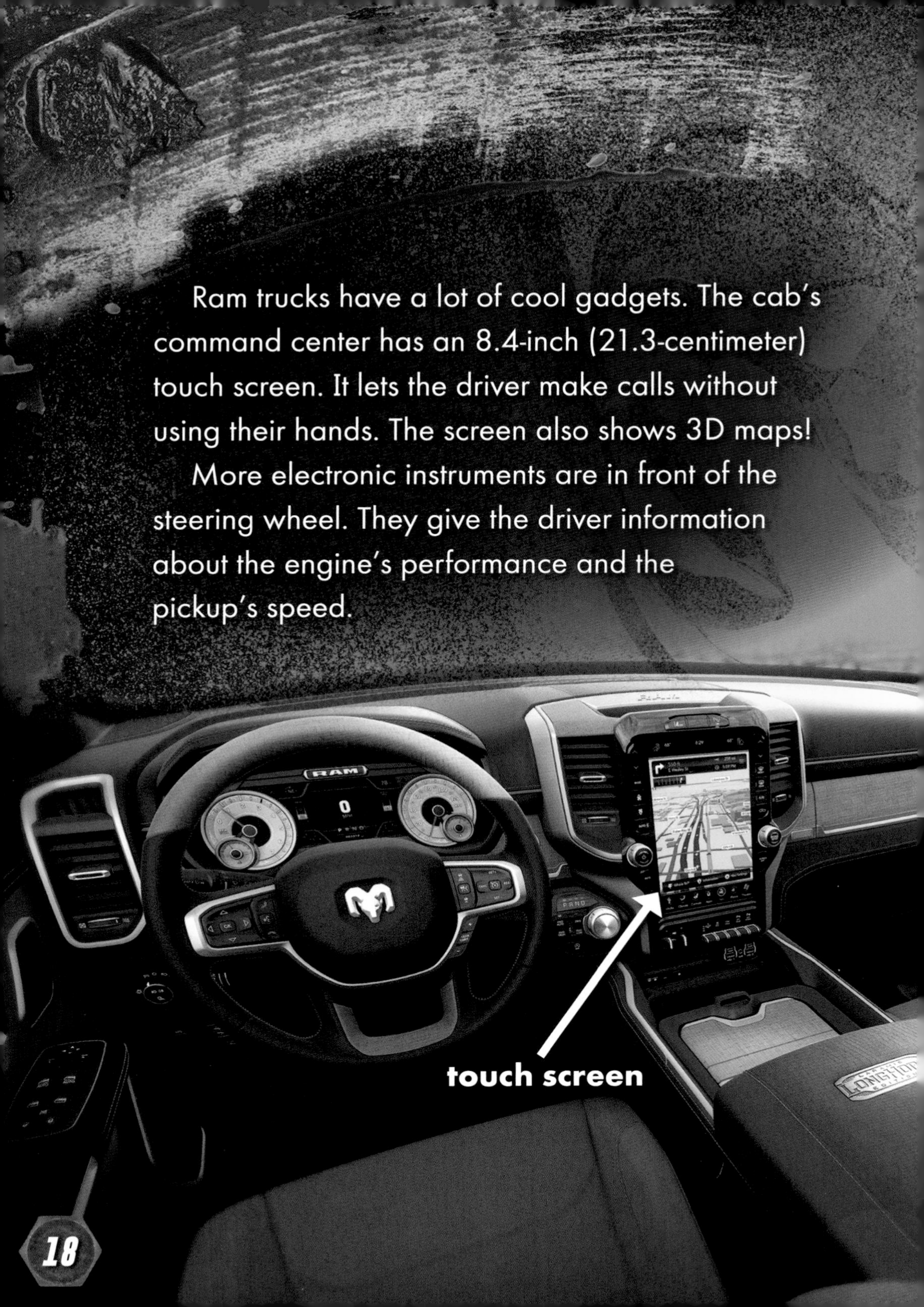

Ram trucks have a lot of cool gadgets. The cab's command center has an 8.4-inch (21.3-centimeter) touch screen. It lets the driver make calls without using their hands. The screen also shows 3D maps!

More electronic instruments are in front of the steering wheel. They give the driver information about the engine's performance and the pickup's speed.

2018 RAM 1500 BIG HORN SPECIFICATIONS

ENGINE	3.6L V6 ENGINE
HORSEPOWER	305 HP (227 KILOWATTS) @ 6,400 RPM
TORQUE	269 LB-FT (37 KG-M) @ 4,175 RPM
TOWING CAPACITY	10,340 POUNDS (4,690 KILOGRAMS)
MAXIMUM PAYLOAD	1,880 POUNDS (853 KILOGRAMS)
FUEL ECONOMY	17 TO 25 MILES PER GALLON
CURB WEIGHT	UP TO 5,512 POUNDS (2,500 KILOGRAMS)
WHEEL SIZE	20 INCHES (51 CENTIMETERS)

TRUCK OF THE FUTURE

Ram trucks have come a long way since the Dodge brothers first started the company. More improvements are planned in the future.

Some Ram 1500s could have **hybrid** engines. These engines will give the trucks even more power. Strength, comfort, and style will bring this powerful pickup even more fans in the future!

GLOSSARY

air springs—parts of a suspension system that release and fill with air to raise and lower the vehicle

brand—a category of products all made by the same company

cab—the area of a pickup where the driver and passengers sit

cylinders—chambers in an engine in which fuel is ignited

division—one part of a company often formed to create a specific product

four-wheel drive—a feature that allows the engine to turn all four wheels at once

hood ornament—a decorative item like a small statue on the front of a vehicle's hood

hybrid—able to use both gasoline and electricity for fuel

muscle cars—high-performance sports cars with strong engines

suspension system—a series of springs and shocks that help a truck grip the road

V8 engine—an engine with 8 cylinders arranged in the shape of a "V"

TO LEARN MORE

AT THE LIBRARY

Bowman, Chris. *Pickup Trucks.* Minneapolis, Minn.: Bellwether Media, 2018.

Mack, Larry. *Chevrolet Silverado.* Minneapolis, Minn.: Bellwether Media, 2019.

Mack, Larry. *Ford F-150.* Minneapolis, Minn.: Bellwether Media, 2019.

ON THE WEB

Learning more about the Ram 1500 is as easy as 1, 2, 3.

1. Go to www.factsurfer.com.

2. Enter "Ram 1500" into the search box.

3. Click the "Surf" button and you will see a list of related web sites.

With factsurfer.com, finding more information is just a click away.

INDEX

The images in this book are reproduced through the courtesy of: meowKa, front cover (hero); Andrey Kuzmin, front cover (title texture), pp. 11 (metal), 14 (metal), 17 (metal), 19 (metal); rootstock, front cover (background); xpixe, front cover (top mud, bottom mud); diogoppr, front cover (mud splash); FCA US MEDIA, pp. 2-3 (logo), 4-5, 6-7, 8, 9, 13, 15, 16, 18, 19, 20, 21 (left, middle, right); Yulia Plekhanova, pp. 2-3 (background); Ed Aldridge, pp. 2-3 (truck); dave_7/ Flickr, p. 10; Darren Brode, p. 11 (hood ornament); Steve Lagreca/ Alamy, p. 12; Gene Blevins/ ZUMAPress, p. 12 (award); Greg Gjerdingen/ Flickr, p. 14; sociologas, p. 21 (metal).